Hydrogeochemi

Y. Sudharshan Reddy
Sunitha Vangala

Hydrogeochemical Studies of Groundwater

in Chennur Mandal, Y.S.R District, A.P using Remote Sensing Technique

LAP LAMBERT Academic Publishing

Imprint

Any brand names and product names mentioned in this book are subject to trademark, brand or patent protection and are trademarks or registered trademarks of their respective holders. The use of brand names, product names, common names, trade names, product descriptions etc. even without a particular marking in this work is in no way to be construed to mean that such names may be regarded as unrestricted in respect of trademark and brand protection legislation and could thus be used by anyone.

Cover image: www.ingimage.com

Publisher:
LAP LAMBERT Academic Publishing
is a trademark of
International Book Market Service Ltd., member of OmniScriptum Publishing Group
17 Meldrum Street, Beau Bassin 71504, Mauritius

Printed at: see last page
ISBN: 978-620-0-50462-3

Table of Contents

ABSTRACT

Water is a non-renewable and most precious resource. Groundwater occupies a major part in water supply for drinking, irrigation and industrial purposes in the world. In India arid and semiarid regions, especially in rural regions, most population depends on the groundwater for their drinking and irrigation purposes. It is evaluated that nearly one third of the world's population depends on groundwater for drinking purpose. Drinking water is an important resource that needs to be protected from pollution and biological contamination. Water is vital to health, well-being, food security and socioeconomic development of mankind. Underground water is clean but it depends upon quality and quantity of minerals dispersed and dissolved in it. Therefore, the presence of contaminants in natural fresh water continues to be one of the most important environmental issues in many areas of the world, particularly in developing countries where several communities are far away from potable water supply. Low-income communities, which rely on untreated surface water and groundwater supplies for domestic and agricultural uses are the most exposed to the impact of poor water quality.

Rural, urban areas depend entirely on groundwater for drinking and around 90% of the rural populace depends on groundwater for domestic purposes (Sunitha et al. 2012). Around the globe nearly 1.5 billion people depend on groundwater for irrigation and drinking purpose. Lithology, geochemistry can alter ground water hydro chemical features further through subsurface movement groundwater chemistry can also be altered through different anthropogenic sources like waste disposal practices, hygienic conditions.

A Geographic Information System (GIS) can be used effectively for this purpose to combine different hydrogeological themes objectively and analyse those systematically for demarcating the potential zone The GIS technology provides suitable alternatives for efficient management of large and complex databases. Watershed monitoring and management has been found to be economical and faster with the usage of the capabilities of a GIS. One of the greatest advantages of using Remote sensing data for hydro geological investigation and monitoring is its ability to generate information in spatial and temporal domain, which is very crucial for successful analysis, prediction and validation Integration of remote sensing data in GIS environment is very useful in delineating various groundwater potential zone is a meaningful way.

The aim of this study is to conduct an assessment of the Hydrogeochemical Studies of Chennur Mandal, Kadapa District, Andhra Pradesh, using Remote Sensing Techniques. The proposed study area in the Chennur Mandal of Kadapa District. The study area falls in the survey of India Top sheet No:57J/14 & 15. Twenty Seven samples of ground water used for drinking purposed were collected from either hand pumps or open wells at different villages of Chennur Mandal of Kadapa District, during the summer season month of March and April 2015. The pH of ground water in the study area is ranging from 7.7 to 8.8. The total hardness of the groundwater in the study area is ranging from 12 to 78 mg/l. Water hardness is primarily due to the result of interaction between water and the Geological formation. The calcium concentration of groundwater in the study area is ranging from 8 to 36 mg/l during the pre-monsoon period. The chloride concentration of the ground water in the study area ranging from 16.9 to 232.2 mg/l during pre-monsoon period. The bicarbonate concentration of the groundwater in the study area in ranging from 48.8 to 280.6 mg/l during the pre-monsoon period.

The fluoride concentration of the groundwater in the study area ranging from 0.78 to 2.62 mg/l during the pre-monsoon period. Low concentration of fluoride (0.78 mg/l) is observed in Nazeerbegpalle village and high concentration of fluoride (2.62 mg/l) is observed Ramanapalle village. Proper deflouridation techniques have to be followed to monitor fluoride contamination.

CHAPTER I

INTRODUCTION

Water is an essential natural resource which is usually abundant for sustaining life and environment. The most obvious medium in which to assess, monitor and control metal pollution is water. Water is known for its high dissolving capacity and is known as the universal solvent. Water is a solvent and also dissolves minerals from the rocks with which it comes in contact. Groundwater, due to its relative purity is well known as the potable water source the world over since time immemorial. The other merits of groundwater over surface water are little or no need of treatment to safeguard its quality, the immense storage potential of aquifers, annual replenishment by rainfall and decentralized facility of taping close to demand centers saving the cost of long transmission. Drinking water is an important resource that needs to be protected from pollution and biological contamination. Underground water is clean but it depends upon quality and quantity of materials dispersed and dissolved in it. Water picks up impurities in during its flow, which are harmful to man and vegetation. The reason for contamination and pollution of water in the natural surroundings and in the storage are pesticides, fertilizers, industrial wastes, inorganic and organic salts from top soil and geological strata. The domestic water bodies are being used for cattle drinking, human bathing, cloths washing and other domestic purposes. The quality of groundwater is highly related with local environmental and geological conditions. The quality of soil and rock and the water table determines the quality of groundwater. Groundwater constitutes an important source of water for drinking, agriculture and industrial production. The use of groundwater has increased significantly in the last decades due to its widespread occurrence and overall good quality. The contribution from groundwater is vital; because about two billion people depend directly upon aquifers for drinking water, and 40 percent of the world's food is produced by irrigated agriculture that relies largely on groundwater. Despite its importance, contamination from natural, human activities, steady increase in demand for water due to rising population and per capita use, increasing need for irrigation, changes in climates and overexploitation etc., among others has affected the use of groundwater as source of drinking water.

Groundwater is an important medium for assessment of pollution and also plays a very important role in almost all the food chains. Most of the groundwater quality problems are due to contamination and over exploitation. Pollutants are added to the groundwater system through natural and anthropogenic activities. The quality of groundwater gradually changes with respect to depth and due to seasonal changes the range and chemical composition of dissolved solids and type of rock that is present in that area. The groundwater quality is the composition of water changes due to physical, chemical, biological and radiological activities. The ground water quality changes due to streambed, seepage and rock sediments.

The quality of groundwater is obtained due to reactions and processes that have been takes place in between the materials that are present on and within the earth's surface and water. The kind and concentration of dissolved solids depend on the source of salts and subsurface environment. The groundwater quality is very significant for mankind as it is directly associated with human race. Quality of groundwater is as important as the quantity. Groundwater quality data provides us the past geological history of the rocks, groundwater recharge, discharge, movement and storage.

Groundwater is an important water resource in India for domestic, agricultural practices and industrial purposes. Now days the demand for fresh water is increasing due to increase of population, urbanization, industrialization and intense agricultural activities in many parts of the world. Unjustified, unreasonable groundwater development and management in the country owing to urbanization, industrialization ever increasing population demands for food security is changing hydrological and geochemical environment of aquifers causing groundwater pollution. Groundwater forms a major source for drinking water in urban as well as in rural areas. Groundwater has the properties of dissolving and carrying in solution, a variety of chemical and other materials. More than 90% of the rural population uses groundwater for domestic purposes (Sunitha V et al., 2012). The quality of ground water is of great importance in determining the suitability of ground water for a certain use (public water supply, irrigation, industrial applications etc). The water used for drinking purpose should be free from any toxic elements, living and nonliving organisms and excessive amount of minerals that may be hazardous to health (Reddy B.M et al., 2013; Sudharshan 2018).

Groundwater is one of the most valuable natural resources, which supports human health, economic development and ecological diversity. Because of its several inherent qualities it has become an immensely important and dependable source of water supplies in all climatic regions including both urban and rural areas of developed and developing countries. Water is important source for life and developmental activity and groundwater is the only reliable resource due to dwindling of surface water sources (Sunitha V et al., 2012a). Providing safe drinking water to rural population is a hectic task for many developing and under developing countries. Since last few decades millions of hand pumps were installed all over India to provide bacteriologically safe drinking water. Groundwater can have some dissolved chemical constituents which may cause chronic health effects (Sunitha V. et al., 2013).

Groundwater is a vital resource for most developmental activities. Demand for groundwater is increasing due to paucity of surface water and recurrent failures of monsoons. Increasing demand for groundwater causes water level to decline and water quality to deteriorate. The quality of drinking water has increasingly been questioned from health point of view for many decades. The severity of environmental problems related to groundwater varies from place to place depending on the geology, hydrologic, climatic conditions and geochemical factors that influence. Moreover, water has been viewed as an infinite and bountiful resource; water today defines human, social and economic development. The alarming rate of population growth, evolving industrial society, advances in technology, and the existing trend of depletion of groundwater resource has raised some serious environmental problems. However, groundwater quality studies with reference to drinking and irrigation purposes in different regions have been carried out, viz., Anantapur (Sunitha et al., 2014; 2012, Kadapa (Sunitha et al., 2016; 2013), Uttarakhand (Jain et al. 2010), Punjab (Kumar et al., 2007). Groundwater in India accounts for about 80% of domestic water requirements and more than 45% of total agricultural water, irrigating 39 million hectares. Groundwater is also the single largest and most productive source of water for irrigation in India. Earlier studies have identified different types of groundwater pollution like geogenic or anthropogenic source, major ions, nitrogen pollutants, trace elements, agricultural pesticides, dissolution or weathering of the rocks and soil.

Water is absolutely necessary natural resource on earth. Safe drinking water is the primary need of every human and also their basic fundamental right. Fresh water has become a scarce commodity due to over exploitation and water is getting contaminated. Groundwater is the most important source of water supply for drinking, agriculture, and industrial purposes. Reliable and accessibility to safe drinking water are needed for sustainable development. Water reliability for different purposes relies on the chemical and physical quality of water. Groundwater chemistry is governed primarily through both natural and anthropogenic variables. The lithological, geochemical composition was impacted by groundwater's hydrochemical features through subsurface movement (Elango et al., 2003). This underground water through pore spaces and weathered formations can change by the action of various different hydrochemical mechanisms (Rajmohan and Elango 2004; Brindah et al., 2012). Groundwater chemistry can be altered through different anthropogenic sources which include point sources like waste disposal practices, hygienic conditions (Amoako et al., 2011; Brindha and Elango, 2015). Efficient evaluation of the physico-chemical parameters, their sources, the control of hydrogeochemical procedures are crucial for the sustainability of the ecosystem. Many scientists have researched physicochemical and hydrochemical parameters to evaluate the groundwater characteristics (Sanchez Martos et al., 1999; Subbarao et al., 2002; Bharadwaj and Singh 2011). Groundwater is considered as a primary requirement owing to the unavailability of a proper source of water. Hence groundwater quality has changed as one of the most crucial environmental issue (Ravichandra and Chandana, 2006). Groundwater is the prime source of drinking water supply for both urban & rural population in India. Groundwater constitutes above 88% of safe drinking water in rural areas (Jain et al., 2010). Apart from drinking, groundwater is also used for irrigation & industrial processes. Advanced methodologies in pumping techniques have made a demand on groundwater in present days. As the surface water pollution is increasing rapidly the extraction of groundwater is also increasing considerably and further groundwater pollution & its misuse is becoming more significant. As people depend on groundwater for various purposes, public health directly linked with quality of drinking water. Its systematic monitoring is very essential. Increased irrigational practices may also lead to groundwater pollution (Pawar and Shaikh 1995; Sunitha et al., 2014). In India groundwater contributes nearly 80% of domestic practices and more than 45% of total agricultural water irrigating 39 million hectares (Kumar et al., 2005).In India groundwater is considered as the only largest & vital viable sources of water for irrigation practices (Kinzelbach et al., 2003). According to the previous literature groundwater pollution is associated with many sources like geogenic or anthropogenic, nitrogen pollutions agricultural practices dissolution of rocks and soil (Guo et al., 2007; Brindha and Elango 2010; Elango et al., 2012; Brindha and Elango 2014; Sunitha et al., 2012; Sunitha et al., 2014; Adimilla and Venkatayogi 2017; Sunitha et al., 2018; Prasad et al., 2019). Groundwater is the prime sources of water for drinking, agricultural needs semi-arid region of Y.S.R District. Geochemistry of groundwater & its sustainability for drinking, agricultural purpose hasn't yet studied hence this study was emphasized on an evaluation of suitability of water for drinking and irrigation practices.

Multidisciplinary scientific integrate surveys were generally carried out to quantify the resource potential of the area, to know the status of exploitation of resources and to identify any degradation due to unscientific management. The investigation agents broadly outline the development options based on available resources. The thematic maps produced on resources will enable planners to formulate programme to optimize productivity from existing resources, and to initiate measures to correct imbalances due to unscientific management and inherent deficiency. Environmental mapping and resource evaluation survey of Chennur Mandal of Kadapa District is taken up identification of areas for further development. Analysis of remotely sensed data for drainage, geological, geomorphological and lineament characteristics of terrain in an integrated way facilitates effective evaluation of ground water potential zones. Similar attempts have been made in the generation of different thematic maps for the delineation of ground potential zones in different part of the study area. (Obi Reddy et al., 1994; Krishna Murthy and Srinivas, 1995; Rao et al; 1996). A total of three thematic maps such as geological, geomorphological and hydrological maps were prepared based on image interpretation studies with limited field checks and analysis of available database (Fig.2,3&4). The lithological map portays distribution of several of rock types and structural maps shows the structural frame work of the area. The geomorphology map depicts the various landforms evaluate through timely by geomorphic process and is a basic input to evaluate resource potential associated with the landforms. The hydrological map provides a basis for potential and non potential areas for groundwater development based on geomorphological, geological and structural information.

Objectives

➢ The present study aims to generate different thematic maps using satellite data along the ancillary data (Geology, Geomorphology, and Geohydrology).

➢ To prepare action plan for water resources

➢ Assessment of water quality by studying hydrogeochemistry.

CHAPTER II

STUDY AREA

The climate of the study area is hot and semiarid. The monthly maximum, minimum and mean temperature as measured at Kadapa are 44°C, 14°C and 27°C respectively. The mean annual rainfall recorded at the Kadapa is 759 mm. The Kadapa district is aptly called the district of Pennar as almost the entire district is drained by the Pennar River and its tributaries. The important tributaries joining the river from the north include the rivers Kunderu, Sagileru and Tummalavanka while those from the south include the rivers Chitravati, Papaghni, Buggavanka, Cheyyeru, and Kalletivagu. Bahuda, Mandavi, Pukkangi and Gunjaneru are the tributaries of the Cheyyeru. The rivers and streams in the district are mostly ephemeral under the influence of heavy spells of rainfall by cyclonic storms in the Bay of Bengal (MRK Reddy et al., 2000). The study area falls in the Survey of India Toposheet No: 57 J/14and J/15. Sample location of the study area is shown in figure 1.

Geology

The oldest rocks of the area belong to Late Archaean or Early Proterozoic era which are succeeded by rocks of Dharwarian Age and both are traversed by dolerite dykes (Murthy et al., 1979). The older rocks are overlain by rocks of Cuddapah Supergroup and Kurnool Group belonging to Middle and Upper Proterozoic Age. The main lithologic units consist chiefly of quartzite, limestone, and shale. Alluvium consisting of gravel, sand, silt and clay occur along the river courses in the study area.

GROUP	FORMATION	LITHOLOGY
	Nandyal Shales	Shale
	Koilkuntla Limestone	Limestone
Kurnool Group	Panyam quartzite	Quartzite

--Unconformity-----------------------------

Cumbum Shales   dolomite   Mud stone/shale

Gulcheru quartzite/conglomerate

--Unconformity-----------------------------

PENINSULAR GNEISSIC COMPLEX Granite Gneisses, Schist, Granitoids with acidic and basic intrusive

Cuddapah Supper Group

The Cuddapah Super Group is represented by thick sequence of sedimentaries unconformabally overlain by the place or basement complex. In the study area Cuddapah Super Group is represented by rock types belonging to Nallamalai group covering an area about 160 sq km.

Nallamali Group

This group is the lower most of the Cuddapah Super Group of rocks and has been divided into two mapable formations viz. Gulcheru formation and Cumbum formation.

Basic intrusives

These dykes are generally medium grained and consists mainly of plagioclase, pyroxenes. Field evidence shows that they are of two generation of dykes. The spectral characteristics of these litho units are tone, texture and linear ridge. These dykes are easily delineated during interpretation.

Geological Structures

Bedding, joints, faults, lineaments, folds, fractures are some of the structure elements interpreted using satellite imagery No: 57 J/14, and J/15. Dykes and faults, Lineaments are the most important structures developed in the area. The lineament either coincide with the drainage directions, alignment with the tanks, vegetation etc.

Bedding

Bedding is manifested by colour banding or compositional layering as observed the formation. The trend of the bedding varies from NW-SE to NNW-SSE with shallow dips (8^0-15^0) two wards NE or ENE.

Lineaments

In the central part of area a major lineament is abuting from balasingayapalle to kanuparthi. Another lineament is extending from Doulathapuram to Shivapalle. Quartz reefs is observed in this lineament.

Faults

The faults are manifested either dykes or displacement of the litho unit or Gulcheru quartzite, and Cumbum shale. Another major fault is EW trending, extends from Pushpagiri to Chennur.

Joints

Occurrence of Joints is prevalent in the quarzites of Gulcheru quarzites in the Balasingayapalle, Kanupahti. The trend of joints varies from SW, these are observed in these villages.

Geomorphology

Geomorphology involves study of landforms, reconstruction of process responsible for their origin and study of influence of tectonics in time space frame. The geomorphological mapping includes inventory and classification of landforms. Each landform depend by its composition depth of weathering structural frame and the environment which includes soil cover, hydrology and hydrogeology. The landforms are classified on the basis of mode of origin, relief slope factor and surface cover. The landforms occurring in the area as grouped as denudational hill, residual hill, pediment, pediplain, cuesta, structural hill, structural valley, and linear ridge.

Residual Hill

In Southern part of the area around Pushpagiri, Kanuparthi isolated low relief hills around denudational hills have been demarcated. A number of linear basic dyke ridges evolved by difference erosional criss cross, the area in EW, NNE-SSW, ENE-SWS directions. Most of these ridges support scrubby vegetation.

Pediment

The pediment shows 2^0-5^0 slope and associated with foot slope element of denudational hill, residual hill and ridges. The pediment is weathered considerably to the depth of few cm to 5 m. Thin coarse red and brownish sandy soils and skeletal gritty soil covered the pediments due to intensive sheet wash process operating in pediment area. The natural vegetation of the pediment comprise open scrub and grass land at places the pediments are under dry condition and groundwater condition is low to moderate.

Pediplain

It is gently sloping ramp of 2-5^0 slope originated by coalescence of several pediment and finally merges with major tributary stream valley and floodplain.

Hydromorphology

Ground water occurrence in hard rock terrain is confined to certain landform and fractures. As the aquifer material and alluvium is usually confined to certain landform. Further lineaments, landform development and their elevation and their elevation and distribution is controlled by faults, streams, segments and fractures. Structurally exclusive litho units like dykes and acidic intrusive, it is evident from the above factors it is imperative that detail landform mapping cum classification elevation and an understanding of morpho techniques is imperative for ground water exploration in hard rock terrain. It is a map which depicts various aspects of geomorphology, geology and character of aquifers so as to have an idea of the possibility of ground water in different units. The hydromorphologic map is to be prepared by demarcating the geomorphic units as the landforms as an important input for land management, soil mapping and identification of potential zones of ground water occurrence. The geological details like lithology, rock types and structural details are also depicted on this map since this information is necessary in identifying the ground water potential. For instance pediment, pediplain without fractures, joints and lineaments normally moderate to poor ground water prospect whereas the same geomorphic unit with a network of fractures, joints indicate good ground water prospects. Similarly pediplain area of crystalline/metamorphic rock is marked by poor to moderate ground water prospect whereas the same unit in sandstone or limestone sedimentary rock may have a good to moderate prospect.

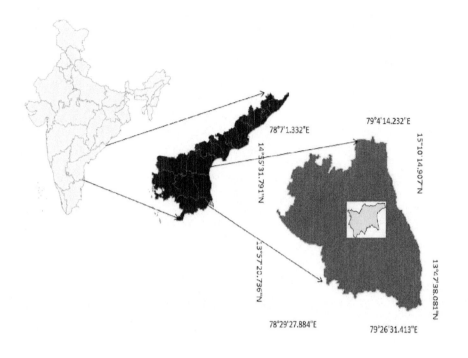

Fig: 1 Location map of the study area

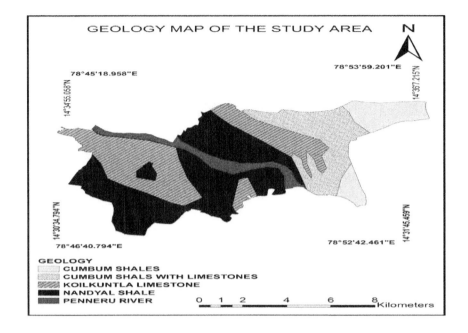

Fig: 2 Geology of the study area

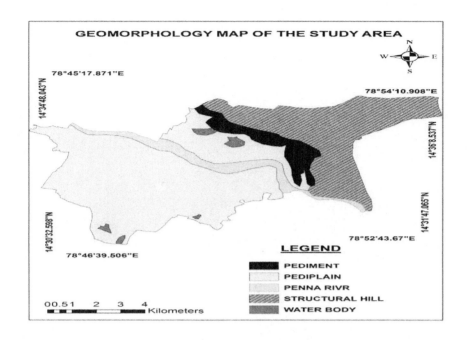

Fig: 3 Geomorphology of the study area

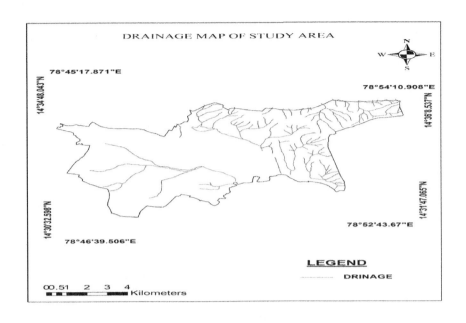

Fig: 4 drainage map of the study area

CHAPTER III

METHODOLOGY

Twenty seven samples are collected in pre monsoon during September 2018. The samples are collected at different villages of Chennur Mandal of Kadapa District during post-monsoon season in the year 2018 in pre-cleaned polyethylene bottles of two litre capacity. This season was selected because in this season often contamination increases due to low dilution and this tends to the accumulation of ions. Before sampling, the water left to run from the source for few minutes. Then water samples collected in pre cleaned, sterilized polyethylene bottles of one litre capacity. The samples were analyzed to assess various physicochemical parameters according to APHA (2007). The water is left to run from the source for about 4 min to stabilize the electrical conductivity (Khaiwal and Garg, 2006). The groundwater samples are analyzed as described by American Public Health Association (APHA 2007) procedure, and suggested precautions are taken to avoid contamination. The various parameters determined are pH, electrical conductivity, total dissolved solids, total hardness, calcium, magnesium, total alkalinity, carbonate, bicarbonate, chloride, sulfate, sodium, potassium, nitrate and fluoride. pH and EC are determined by pH meter, conductivity meter, TDS are determined by indirect method (Raghunath, 2003), Total Hardness, Ca^{2+}, Mg^{2+}, CO_3^{2-}, HCO_3^- and Cl^- are determined by titrimetry, whereas Na^+ and K^+ are determined by flame photometry (Systronic Model No.128), SO_4^{2-} and NO_3^- are determined by spectrophotometric method. F^- is determined by using ion selective electrode (Orion 4 star ion meter, Model: pH/ISE). The depth of the bore wells varied between 250 and 700 feet.

CHAPTER IV

RESULTS AND DISCUSSION

Water Quality

The concept of water quality is complex because so many factors influence in it. In particular, this concept is intrinsically tied to the different intended uses of the water; different uses require different criteria. Water quality is one of the most important factors that must be considered when evaluating the sustainable development of a given region (Cordoba et al., 2010). Water quality must be defined based on a set of physical and chemical variables that are closely related to the water's intended use. For each variable, acceptable and unacceptable values must then be defined. Water whose variables meet the pre-established standards for a given use is considered suitable for that use. If the water fills to meet these standards, it must be treated before use. Water quality is considered the main factor controlling health and the state of disease in both man and animals.

Water is vital to health, well-being, food security and socioeconomic development of mankind. Therefore, the presence of contaminants in natural freshwater continues to be one of the most important environmental issues in many areas of the world, particularly in developing countries, where several communities are far away from potable water supply. Low-income communities, which rely on untreated surface water and groundwater supplies for domestic and agricultural uses are the most exposed to the impact of poor water quality. Unfortunately, they are also the ones that do not have adequate infrastructure to monitor water quality regularly and implement control strategies. (Ayoko et al., 2007; Kazi et al., 2009) reported that human activities are a major factor determining the quality of the surface and groundwater through atmospheric pollution, effluent discharges, use of agricultural chemicals, eroded soils and land use. Environmental pollution, mainly of water sources, has become public interest. The chemical composition of ground water is controlled by many factors that include the composition of precipitation, mineralogy of the watershed and aquifers, climate and topography. These factors can combine to create diverse water types that change in composition spatially and temporally (Chenini I and Khemiri S., 2009). Exploitation of groundwater resources beyond their potential renewal capacity, results in a hydrological deficit. Generally, this is expressed as a decline in groundwater levels but in coastal aquifers this may cause intrusion of seawater.

Concept of Ground Water Quality

The concept of ground water quality seems to be clear, but the way of how to study and evaluate it still remains tricky (Chenini I and Khemiri S., 2009) (Badiker et al.,2007). Consider that the definition of water quality is not objective, but is socially defined depending on the desired use of water. Different uses require different standards of water quality.

Safe Drinking water

Potable or "drinking" water can be defined as the water delivered to the consumer that can be not used for drinking, cooking. Only used washing. This water must meet the physical, chemical, bacteriological and radionuclide parameters when supplied by an approved source, delivered to the consumer through a protected distribution system in sufficient quantity and pressure. (Zuane J., 1997).

Water Quality Standards/Guidelines: The Guidelines for drinking-water quality

The Guidelines describe reasonable minimum requirements of safe practice to protect the health of consumes and/or derive numerical "guideline values" for constituents of water or indicators of water quality. In order to define mandatory limits, it is preferable to consider the guidelines in the context of local or national environmental, social, economic and cultural conditions (WHO, 2008).

The Standard for drinking-water quality

By definition, a standard "a rule or principle considered by an authority and by general consent as a basis of comparison. It is something normal or average in quality and the most common form of its kind". A proper standard for drinking water quality is thus the reference that will ensure that the water will not be harmful to human health. The framework against which a water sample can be considered good or "safe" is a drinking water quality standard (Solsona F, 2002).

WHO Guidelines

The primary purpose of the Guidelines for Drinking-water Quality is the protection of public health. Water is essential to sustain life, and a satisfactory (adequate, safe and accessible) supply must be available to all improving access to safe drinking-water (WHO, 2008).

Field Photographs

Fig: 5 Sample Collections at Chennur Madal, Y.S.R District

S No	pH	EC (µs)	TDS (mg/L)	Chloride (mg/L)	Total Hardness (mg/L)	Fluoride (mg/L)	Carbonate (mg/L)	carbonate (mg/L)	Calcium (mg/L)
1	8.32	3740	1630	93.7	72	1.09	24	122	36
2	8.16	2920	1220	53.3	48	1.97	252	48.8	12
3	8.78	2400	1060	55.4	30	2.61	12	85.4	12
4	8.13	2210	960	46.8	42	0.9	12	73.2	16
5	8.22	2260	990	51.1	36	1.51	12	73.2	20
6	8.06	2310	1010	61.8	54	1.2	6	48.8	12
7	8.12	2620	1160	78.8	78	1.43	12	85.4	20
8	8.14	2690	1170	232.2	24	2.93	12	122	20
9	8.14	3160	1430	72.4	42	1.35	198	85.4	24
10	8.37	6200	2760	102.7	54	1.03	60	48.8	12
11	8.57	5980	2630	127.8	48	1.27	24	280.6	8
12	8.63	4740	2040	70.3	36	0.774	6	244	16
13	8.42	2310	990	51.1	48	1.35	42	244	12
14	8.06	1640	730	42.6	42	1.38	6	146.4	16
15	8.62	3680	1630	78.1	36	1.11	0	61	20
16	7.86	3150	1340	16.9	66	0.732	6	158.6	32
17	7.96	2540	1120	62.7	72	0.967	6	109.8	28
18	8.08	3740	1660	80.4	72	0.848	6	122	28
19	7.59	4470	1970	90.1	72	0.885	6	122	36
20	7.7	3790	1630	78.1	30	0.963	12	170.8	24
21	7.88	3450	1500	71.2	12	1.3	6	146.4	28
22	7.97	2220	230	35.2	30	0.505	12	73.2	28
23	7.98	2290	990	51.0	54	1.25	12	97.6	36
24	7.81	4980	2170	101.2	54	1.32	6	122	28
25	7.41	3080	1210	79.8	36	0.892	0	48.8	16
26	8.15	4010	1960	89.1	48	2	6	195.2	20
27	8.19	3840	1620	77.2	18	1.54	6	109.8	20

Table: 1 Physico Chemical Parameters of ground water of the study area

SAMPLE NO	NAME OF THE VILLAGE	LATITUDE	LONGITUDE
1	CHINNAMACHUPALLI	$14^013'07"N$	$78^033'52.9"E$
2	CHINNAMACHUPALLI	$14^032'19.8"N$	$78^047'44.7"E$
3	RAMANAPALLI	$14^032'18"N$	$78^048'23"E$
4	RAMANAPALLI	$14^032'17.8"N$	$78^048'44.2"E$
5	RACHINNAYAPALLI	$14^031'36.3"N$	$78^048'48.1"E$
6	RACHINNAYAPALLI	$14^032'19.0"N$	$78^048'47.9"E$
7	MUNDLAPALLI	$14^032'22.9"N$	$78^041'24.5"E$
8	MUNDLAPALLI	$14^032'19.4"N$	$78^049'13.7"E$
9	OBULAMPALLI	$14^032'19.3"N$	$78^049'13.7"E$
10	OBULAMPALLI	$14^032'45.3"N$	$78^050'01.5"E$
11	NAZEERBEGPALLI	$14^032'54.3"N$	$78^049'51.1"E$
12	NAJEERBEGPALLI	$14^032'59.6"N$	$78^049'53.1"E$
13	GURRAMPADU	$14^033'1.8"N$	$78^049'34.4"E$
14	GURRAMPADU	$14^032'59.5"N$	$78^049'33.2"E$
15	CHENNUR CHENNUR	$14^034'30.7"N$	$78^047'38.3"E$
16		$14^034'30.7"N$	$78^047'38.3"E$
17	BALASINGAYAPALLI	$14^03339.6"N$	$78^048'21.2"E$
18	BALASINGAYAPALLI	$14^033'40.4"N$	$78^049'37.1"E$
19	KANUPARTHI KANUPARTHI	$14^033'39.8"N$	$78^049'00.4"E$
20		$14^033'42.5"N$	$78^048'59.2"E$
21	BAYANAPALLE	$14^034'00.0"N$	$78^048'3702"E$
22	BAYANAPALLE	$14^034'20.5"N$	$78^048'43.1"E$
23	DOULATHAPURAM	$14^035'2409"N$	$78^049'01.6"E$
	DOULATHAPURAM	$14^035'24.4"N$	$78^048'26.8"E$
24	KOKKARAYAPALLE	$14^034'36.8"N$	$78^046'57.3"E$
25	SIVALAPALLE	$14^034'27.9"N$	$78^045'58.3"E$
26	PUSHPAGIRI	$14^035'39.9"N$	$78^048'31.9"E$
27	DUGANAPALLE	$14^032'34.3"N$	$78^046'46.6"E$
28	UPPARAPALLE	$14^032'38.7"N$	$78^047'36.6"E$

Table: 2 Water Sample Locations

Salient features of major ion chemistry

pH

The pH of water is very important of its quality and provides important piece of information in many types of geochemical equilibrium or solubility calculations (Hem, 1991). The limit of pH value for drinking water is specified as 6.5 to 8.5 (ISI, 1983). In most natural waters, the pH value is dependent on the carbon dioxide-carbonate-bicarbonate equilibrium. As the equilibrium is markedly affected by temperature and pressure, it is obvious that changes in pH may occur when these are altered. Most ground waters have a pH range of 6 to 8.5 (Karanth, 1987). The pH of groundwater in the study area is ranging from 7.4 to 8.8. pH values for all the samples are within the desirable limits. It is observed that most of the groundwater is alkaline in nature. Though pH has no direct effect on the human health, all biochemical reactions are sensitive to variation of the pH.

Total Hardness

Hardness is an important criterion for determining the usability of water for domestic, drinking and many industrial purposes (Karanth, 1987) and results from the presence of divalent metallic ions, of which calcium and magnesium are the most abundant in the groundwater. Other elements could be included are strontium, barium and some heavy metals. These, however are seldom determined under usually present in insignificant amounts relative to calcium and magnesium

The degree of hardness in water is commonly based on the following classification:

Hardness classification of water	
(After Sawyer and Mc Carty)	
Hardness, mg/l as CaCO₃	**Water class**
0-75	Soft
75-100	Moderately hard
150-300	Hard
Over 300	Very hard

The total hardness of the groundwater in the study area is ranging from 12 to 78 mg/l. The limit of total hardness for drinking water is specified as 300 mg/l (ISI, 1983). Water hardness is primarily due to the result of interaction between water and the geological formations. Groundwater of the entire study area exceed the desirable limits. Granitic rocks significantly contribute to groundwater hardness.

Calcium

The range of calcium content in groundwater is largely dependent on the solubility of calcium carbonate, sulfate and rarely chloride. The solubility of calcium carbonate varies widely with the partial pressure of CO_2 in the air in contact with the water. The salts of calcium and magnesium are responsible for the hardness of water. The permissible limit of calcium in drinking water is 75 mg/l (ISI, 1983). The calcium concentration of the groundwater in the study area is ranging from 8 mg/l to 36 mg/l during pre- monsoon period.

Chloride

Chloride bearing rock minerals such as sodalite and chlorapatite which are very minor constituents of igneous and metamorphic rocks, and liquid inclusions which comprise very insignificant fraction of the rock volume are minor sources are chloride in groundwater. It is presumable that the bulk of the chloride in groundwater is either from atmospheric sources or sea-water contamination. Most chloride in groundwater is present as sodium chloride, but the chloride content may exceed the sodium due to base-exchange phenomena (Karanth, 1985) and also weathering of phosphate minerals and domestic sewage (Karanth, 1987). The upper limit of chloride concentration for drinking water is specified as 250 mg/l (ISI, 1983). The chloride concentration of the groundwater in the study area is ranging from 16.9 to 232.4 mg/l during pre-monsoon period.

Fluoride

Fluoride in drinking water has now become one of the most important geo-environmental and toxicological issues in the world. During the last three decades, high fluoride concentrations in drinking water sources and the resultant disease "Fluorosis" is being highlighted throughout the world. In developing countries, especially in the tropical regions, rural communities, how mostly depend on groundwater sources for their domestic water supplies, face this problem seriously. In India alone about 25 million people in 8700 villages are consuming water with high fluoride concentrations (Apambire et al, 1997). Andhra Pradesh, Rajasthan, Uttar Pradesh, Gujarat and Tamil Nadu being the states with the highest rating of endemic Fluorosis. Fluoride concentrations as much as 20 mg/l have been recorded in groundwater from these areas (Handa, 1975). The prominent health related problems due to high fluoride concentration are dental concentrations as much as 20 mg/l have been recorded in groundwater from these areas (Handa, 1975). The prominent health related problems due to high fluoride concentration are dental caries, teeth mottling endemic cumulative fluorosis causing skeletal damage and deformation to children and adults. According to Indian standard specification for drinking water (BIS, 1991; WHO, 1985), 1.5-mg/l fluoride is the maximum permissible limit. Dental fluorosis is visible sign that fluoride has poisoned enzymes in the body. Poor nutrition exacerbates the toxic effects of fluoride exposure, which is the region why the poor communities are common victims. According to the agency for toxic substances and Disease Registry, "Existing data indicate the subsets of the population include the elderly people with deficiency of calcium, magnesium and vitamin C, and people with cardiovascular and kidney problems".

In recent times the interest in fluoride has greatly increased, owing to its importance in the precipitation of fixation of phosphate in minerals like fluorapatite, and to the recognition of pathological conditions in man and animals, described as fluorosis.

Fig 5: Anaysis of Physico Chemical Paramters in wet chemical laboratory

Effect of Fluoride on human health

Fluoride is beneficial to certain extent when present in concentrations of 0.8-1.0 mg/l for calcification of dental enamel especially for the children below 8 years of age. But it causes dental fluorosis if present in excess of 1.5 mg/l and skeletal fluorosis beyond 3.0 mg/l if such water is consumed for about 8 to 10 years. As per W.H.O guidelines for drinking water quality and water technology mission of the Government of India, the permissible limit for fluoride in drinking water is 1.0 mg/l. It can be extended to 1.5 mg/l if there is no alternative source in the village.

Dental Fluorosis

Generally ingestion of water having a fluoride concentration above 1.5- 2.0 mg/l may lead to dental mottling, an early sign of dental fluorosis which is characterized by opaque white followed by pitting of teeth surfaces. The degree of mottling depends largely on the amount of fluoride ingested per day. With increasing concentrations of fluoride, the effect becomes progressive. Dental fluorosis can produce considerable added dental costs (tooth deterioration) and significant physiological stress for affected populations.

Skeletal Fluorosis

Skeletal fluorosis may occur when fluoride concentrations in drinking water exceed 4-8 mg/l. The condition itself as an increase in bone density leading to thickness of long bones and calcification of ligaments. The disease may be present in an individual at subclinical, chronic, or acute levels of manifestation. Symptoms range from mild rheumatic or arthritic pain in joints and muscles, to severe pain in the cervical spine region along with stiffness and rigidity of the joints. Crippling skeletal fluorosis can occur when a water supply contains more than 10 mg/l (WHO 1970). The fluoride concentration of the groundwater in the study area is ranging from 0.5 mg/l to 3 mg/l during pre-monsoon period. Low concentration of fluoride (0.411 mg/l) is observed in Bayanapalle village and high concentration of fluoride (6.18 mg/l) is observed in the village Mundlapalle.

Total Alkalinity (CO_3 and HCO_3)

The primary source of carbonate and bicarbonate ions in groundwater is the dissolved carbon dioxide in rain, which, as it enters the soil, dissolves more carbon dioxide. An increase in temperature or decrease in the pressure causes reduction in the solubility of carbon dioxide in water (Karanth, 1989). The alkalinity of natural water is due to the salts of carbonates, bicarbonates, borates, silicates and phosphates along with hydroxyl ions in the free salt. However, the major portion of the alkalinity in natural water is caused by hydroxide, carbonate and bicarbonates, which may be ranked inorder of their association with pH values. The bicarbonate concentration of the groundwater in the study area is ranging from 61 mg/l to 280.6 mg/l during pre-monsoon period. The permissible limit of carbonate (CO_3) in drinking water is 10 mg/l and the rejection limit is 50 mg/l. The permissible limit of bicarbonate (HCO_3) in drinking water is 500 mg/l. (Todd, 1980). Most of the water samples of the study area contains no carbonate ions.

CHAPTER V

CONCLUSION

The pH of groundwater in the study area is ranging from 7.41 to 8.78. pH values for all the samples are within the desirable limits. It is observed that most of the groundwater is alkaline in nature. The electrical conductivity of the groundwater is ranging from 1640 μSiemens/cm-6200 μSiemens/cm at 25°C. The pH and EC were measured with pH meter and conductivity meter respectively. The Total Hardness of the groundwater in the study area is ranging from 12 to 78 mg/l. The limit of Total Hardness for drinking water is specified as 300 mg/l (ISI, 1983). Water hardness is primarily due to the result of interaction between water and the geological formations. Groundwater of the entire study area exceed the desirable limits. The calcium concentration of the groundwater in the study area is ranging from 8 mg/l to 36 mg/l during pre- monsoon period. The upper limit of chloride concentration for drinking water is specified as 250 mg/l (ISI, 1983). The chloride concentration of the groundwater in the study area in ranging from 16.9 to 232.2 mg/l during pre-monsoon period. The bicarbonate concentration of the groundwater in the study area is ranging from 61 mg/l to 280.6 mg/l during pre-monsoon period. The fluoride concentration of the groundwater in the study area is ranging from 0.5 mg/l to 3 mg/l during pre-monsoon period. Low concentration of fluoride (0.505 mg/l) is observed in Bayanapalle village and high concentration of fluoride (3 mg/l) is observed in the village Mundlapalle. Proper defluoridation techniques have to be followed to monitor fluoride contamination.

REFERENCES

1.   Apambire WB, Boyle DR, Michel FA (1997) Geochemistry, genesis, and health implications of fluoriferous groundwater in the upper regions of Ghana. Environ Geol 33(1):13-24.

2.   APHA (2007) Standard methods for the examination of water and wastewater. American Public Health Association, Washington, DC.

3.   Ayoko G., Singh K., Balerea S., Kokot S. (2007), "Exploratory multivariate modeling and prediction of the physic – chemical properties of surface water and groundwater". Journal of Hydrology 336, 115- 124.

4.   Babiker I., Mohamed M., Hiyama T., (2007) "Assessing groundwater quality using GIS". Water Resources Man agement 21,699 – 715.

5.   BIS (1991), Specification for drinking water, IS: 10500:1991, Bureau of Indian Standards, New Delhi.

6.   Chenini I., Khemiri S., (2009), "Evaluation of ground water quality using multiple linear regression and structural equation modeling". Int. J. Environ. Sci. Tech., 6(3), 509-519.

7.   Frank RG. and Shannon M, (2005). Infant Methemoglobinemia: The Role of Dietary Nitrate in Food and Water. *Pediatrics.* 116:784.

8.   ISI (1983). Drinking water standars, Table 1, Substance and characteristics affecting the acceptability of water for domestic use 18, 10500. Indian Standard Institution, New Delhi.

9.   Todd DK. (1980). Groundwater Hydrology, Wiley-Indian Edition.

10.   Krishnaswamy VS (1981) Geological and mineral map of Cuddapah basin. Geol Surv India.

11.   Karanth   KR   (1989)   Groundwater   assessment,   Development   and Management. Tata McGraw-Hill, New Delhi, p 720.

12.   Karanth, K. R., Jagannathan, V., Prakash, V. S. & Saivasan, V. (1992) Pegmatites: a potential source for sitting high yielding wells. J. Geol. Soc. India 39, 77-81.

13.   Kazi T., Arain M. , Jamali M., Jalbani N. , Afridi H., Sarfraz R ., Baig J., Shah A., (2009) "Assessment of water quality of polluted lake using multivariate statistical techniques: A case study". Ecotoxicology and Environmental Safety 72, 301 – 309.

14.   Morris, B.L. et al. (2003), Groundwater and its susceptibility to degradation: a global assessment of the problem and options for management, Early Warning and Assessment Report Series.

15.   Nanoti M, (2004). "Importance of water quality control in treatment and provision of safe public water supply", *National workshop on control and mitigation of excess fluoride in drinking water,* 5-7th Feb, 2004.

16.   Ramakrishna Reddy M., Janardhana Raju N., Venkatarami Reddy Y., Reddy T.KV.K 2000 Water Resources Development and management in the Cuddapah district, India, Environmental Geology 39(3-4) January.

17.   Solsona F., (2002). "Guidelines for Drinking Water Quality Standards In Developing Countries".

18.   Tambekar DH, Waghode SM, Ingole SG and Gulhane SR, (2008). Water quality index analysis of salinity affected villages from the Purna river basin of Vidarbha region. International Quarterly Scientific Journal Nature Environment and Pollution Technology, 7(4): 707 – 711.

19.   World Health Organization, (2008), "Guidelines for drinking-water quality". Third edition.

20.   Zuane J., (1997), "Handbook of Drinking Water Quality". Second Edition, John Wiley & Sons, Inc.

21.   Adimilla,N., Venkatayogi, S., 2017; Amoako, J., Karikari, AY., Ansa-Asare, OD., 2011. Physico-chemical quality of boreholes in Densu Basin of Ghana. Appl Water Sci. 1:41–48

22.   Amoako, J., Karikari, AY., Ansa-Asare, OD., 2011. Physico-chemical quality of boreholes in Densu Basin of Ghana. Appl Water Sci 1:41–48

23.   Brindha, K., Elango, L., 2015. Cross comparison of five popular groundwater pollution vulnerability index approaches, Journal of Hydrology (2015), doi: http://dx.doi.org/10.1016/j.jhydrol.2015.03.003

24.   Brindha, K., Elango, L., 2014. Spatial analysis of soil fertility in a part of Nalgonda district, Andhra Pradesh, India. Earth Science India. 7(1), 36-48

25.   Brindha, K., Elango, L., 2012.   Impact of Tanning Industries on Groundwater Quality near a Metropolitan City in India Water Resour Manage. 26:1747–1761.DOI 10.1007/s11269-012-9985-4.

26.   Brindha, K., Elango, L., 2010. Study on bromide in groundwater in parts of Nalgonda district, Andhra Pradesh. Earth Science India. 3(1), pp. 73-80

27.   Bhardwaj, V., Singh, DS., 2011. Surface and groundwater quality characterization of Deoria District, Ganga Plain, India. Environ Earth Sci. 63:383–395.

28.   Elango, L., Brindha, K., Kalpana, L., Faby Sunny Nair, R. N., Murugan, R., 2012 .Groundwater flow and radionuclide decay-chain transport modeling around a proposed uranium tailings pond in India. Hydrogeology Journal. 20: 797–812. DOI 10.1007/s10040-012-0834-6.

29.   Elango, L., Kannan, R., Senthil Kumar, M., 2003. Major ion chemistry and identification of hydrogeochemical processes of groundwater in a part of Kancheepuram district, Tamil Nadu, India. J Environ Geosci.10:157–166.

30.   Guo, F., Jiang, G., Yuan, D., 2007. Major ions in typical subterranean rivers and their anthropogenic impacts in southwest karst areas, China. Environ Geol. 53:533–541

31.   Jain, C.K., Bandyopadhyay, A., Bhadra, A., 2010. Assessment of ground water quality for drinking purpose, District Nainital, Uttarakhand, India. Environ Monit Assess. 166(1–4):663–676.

32.   Kinzelbach, W., Bauer, P., Siegfried, T., Brunner, P., 2003. Sustainable groundwater management—problems and scientific tools. 26(4), 279–284.

33.   Kumar, R., Singh, RD., Sharma, KD., 2005. Water resources of India. Curr Sci. 89(5), 794–811

34.   Prasad, M., Sunitha, V., Sudharsan Reddy, Y., Suvarna, B., Muralidhar Reddy, B., Ramakrishna Reddy, M; 2019. Data on Water quality index development for groundwater quality assessment from Obulavaripalli Mandal, YSR district, A.P India. Data in brief. https://doi.org/10.1016/j.dib.2019.103846.

35.   Pawar, NJ., Shaikh. IJ., 1995. Nitrate pollution of ground waters from shallow basaltic aquifers, Deccan trap Hydrologic Province, India. Environ Geol. 25:197–204.

36.   Rajmohan, N., Elango, L., 2004. Identification and evolution of hydrogeochemical processes in the groundwater environment in an area of the Palar and Cheyyar River Basins, Southern India. Environ Geol. 46:47–61.

37.   Ravichandra, R., Chandana, O.S., 2006. Study on evaluation on ground water pollution in Bakkannaplem, Visakhapatnam. Nature, Environment and Pollution Technology. 5(2), 203-207.

38.   Reddy Sudarshan, Y., Sunitha, V., Suvarna, B., Prasad, M., 2018. Assessment of Groundwater quality with special reference to Fluoride in groundwater surroundingabandonedmine sites at Vemula, Y.S.R district, A.P. Journal of Emerging Technologies and Innovative Research (JETIR). 5 (11),769-777.

39.   Reddy Muralidhara, B., Sunitha, V., Prasad, M, Sudharshan Reddy, Y., Ramakrishna Reddy, M., 2019. Evaluation of groundwater suitability for domestic and agricultural utility in semi-arid region of Anantapur, Andhra Pradesh State, South India. Groundwater for Sustainable Development 9 100262. Elsevier Journal

https://doi.org/10.1016/j.gsd.2019.100262.

40.   Subba Rao, N., Marghade, D., Dinakar, A., 2017. Geochemical characteristics and controlling factors of chemical composition of groundwater in a part of Guntur district, Andhra Pradesh, India. Environ Earth Sci. 76, 747.

Sanchez Martos, F., Pulido Bosch, A., Calaforra, JM., 1999. Hydrogeochemical processes in an arid region of Europe (Almeria, SE Spain). Appl Geochem.14:735–1.

Printed in Great Britain
by Amazon

82430872R00037